gift of the

*K*athryn
*W*ilson
*R*oberts
fund

The
BEE-KIND
Garden

The
BEE-KIND
Garden

Apian wisdom for your garden

DAVID SQUIRE

green books

CONTENTS

CHAPTER ONE

Getting to Know Bees

CHAPTER TWO

Attracting Bees

INTRODUCTION

Bee-keeping is an ancient craft, and one steeped in tradition and folklore. It has a shepherding nature, with the apiarist caring for the health and welfare of the hive's residents but without being able to impose strict discipline on them. It must be remembered that although bees can be domesticated, they will only remain in a hive if they are treated with respect and given comfortable conditions. In exchange for looking after them and, usually, feeding with a sugary syrup in late summer, autumn and winter, the bees will provide you with honey. It is more than a fair exchange.

The Beekeepers, 1567–68, Pieter Bruegel the Elder

A colony of bees is complex, and its ability to regulate and organize its own affairs is remarkable; all residents, whether the queen, drones or the thousands of worker bees, know their roles and exactly what is expected from them throughout the year.

From *The English Illustrated Magazine*, 1890

Honey bees are reliant on both pollen and nectar, which they gather from native plants as well as from ornamental ones in flower beds and borders. Orchards are another source of pollen and nectar. To ensure good health and production of honey, bees need a wide range of flowers that are, essentially, uncontaminated by chemical sprays. They also need a readily available supply of clean water.

Bee-keeping is an exciting and satisfying hobby; it is practical, fulfilling and contemplative, with the bonus of working in harmony with nature.

CHAPTER ONE

Getting to Know Bees

WHAT ARE BEES?

Honey bees are social insects that have been domesticated and encouraged to live in hives. Earlier, bees made homes in holes and crevices in trees, walls and logs; later they used 'skeps', then other early hives. Now, they live in man-made hives designed and constructed to meet all their needs.

Skep, English woodcut, 1658

WHAT ARE HONEY BEES?

They belong to a classification of insects known as *Hymenoptera*, meaning having membranous wings. Relatives of honey bees include wasps, hornets, bumble bees, ants and sawflies. They all have two pairs of more or less transparent wings.

Honey bees

HYMENOPTERA INSECTS

These are divided into two groups. The first includes sawflies and their close associates. They do not have a constricted waist, and the egg-laying part, known as an ovipositor, is modified to form a saw. This enables females to saw slits or pockets into which eggs can be laid. The other group has a constricted waist, and includes bees and wasps. Ants are also in this group, but they have mostly lost their wings (except for sexually active males and females).

STINGING OR LAYING EGGS?

Two groups of insects within the *Hymenoptera* have an ovipositor or a modification of it. With ichneumon and chalcid wasps this has been adapted to lay eggs on or within the bodies of insects which later serve as hosts and food reservoirs for their offspring. The ovipositor in bees and wasps has become modified into a sting. Female bees, wasps and ants, however, are able to discharge eggs through an opening at the base of their ovipositor.

1 Winged ant. 2 Hornet. 3 Wasp. 4 Ichneumon fly.

LIFE SPANS

Life expectancies of a hive's occupants (queen bee, drones and workers) vary throughout the seasons and especially from winter to summer. Each insect has its own agenda and duties, and they all act in harmony. It is an interactive social existence, and life spans reflect the nature of their roles.

Frontispiece, *Parliament of Bees* (written 1608–16, but not published until 1641) by John Day (1574–1638)

Dutch woodcut, 1488

COLONY HEAD

A queen bee usually lives for two or three years and during this period lays 600,000 eggs. However, records indicate that some queens live for four or five years – or even longer. When her egg-laying abilities decline, her place is taken by one of her daughters, specially reared for this purpose.

A DRONE'S LIFE EXPECTANCY

During summer a drone usually lives for four or five weeks; he becomes sexually active when 10–14 days old, and his main duty is fertilizing the queen. Drones do not visit flowers for food; instead, they either remove it from returning foraging worker bees or take it from stored food within the hive. The thought that drones have a comfortable existence is confirmed by them usually congregating in the warmest part of the hive and only going outside on fine, sunny and windless days – and then only in the afternoon between midday and four o'clock.

WORKER BEES

Their life expectancy varies throughout the year, with those born from late spring to mid-summer (known as 'summer' bees) living for an average of 25 days. However, those born in early autumn (often known as 'winter' bees) live for 50 days or more. This life expectancy is influenced not only by the amount of work expected from them from one season to another (including foraging, looking after young larvae and other domestic duties), but also by an internal change influenced by the number of larvae required to replace them.

A HIVE'S RESIDENTS

The adult residents of hive are a queen, drones and worker bees. Additionally, there are eggs and larvae which, when given the correct attention and food, develop into adults (mainly workers). Each adult bee has an individual role to play in the continuing welfare of a colony.

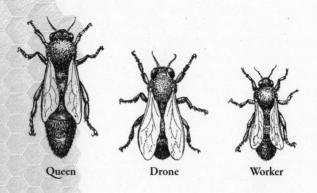

Queen   Drone   Worker

THE QUEEN BEE

In an established hive, there is normally only one queen bee (although slightly prior and just after swarming there would be young queen bees ready to emerge to replace her). She is the largest of a hive's occupants and her only job is to mate with a drone and to lay eggs.

PRODIGIOUS EGG-LAYING

The main task of a queen bee is to lay eggs; during early summer she will lay about 1,500 eggs every day. This means that in every 24 hours she is producing more than her own body weight in eggs.

DRONES

These are male bees, and their numbers vary throughout the year, especially during swarming; usually they form only about 1% (or less) of a hive's occupants. They are smaller than a queen bee, and have an ungainly and round head.

From egg to larva, then adult

WORKER BEES

These are infertile females, smaller than drones, and the essential 'workers' within a hive. They forage for nectar and pollen, fan the hive to keep it cool in summer, act defensively against marauders and look after the queen bee in her vital role of laying eggs. Usually, there are 55,000 or more worker bees in a hive, but this can dramatically increase in large colonies.

LARVAL STAGES

Part of a bee's development is its larval stage. Eggs are laid by the queen bee in hexagonal wax cells created by worker bees. These cells form a comb, and each cell has a slightly downward slope towards its base. Eggs laid in them develop into larvae (resembling small, legless grubs) which later mainly become worker bees.

COLONY REPRODUCTION

When a swarm of bees emerges from a hive, it is nature's way of ensuring the perpetuation of the species. The impulse for the occupants of a hive to swarm is usually because it is congested and more space is needed. Swarming can occur from late spring to mid-summer.

Queen bee and attendants

CREATION OF A REPLACEMENT QUEEN

When the queen decides to leave a hive and create a swarm, worker bees react and start to construct several special cells, which are known as queen cells. The queen lays eggs into around 12 of these cells.

DEVELOPMENT OF QUEENS

The eggs develop, and at their larval stage worker bees lavishly feed them with food secreted by their pharyngeal salivary glands. The larvae develop and grow rapidly, and the cell walls enlarge to accommodate their increasing size.

VIRGIN QUEENS

When a colony swarms, the old queen, together with worker bees and a few drones, leaves the hive and seeks a new home. Between five and eight days after the swarm departs, a virgin queen emerges from her cell and, unless prevented by worker bees, destroys all other virgin queen cells.

THE PRICE OF FATHERHOOD!

Immediately after fertilization, the queen is said to tear herself away from the drone violently, leaving his genitalia still inside her. Alternatively, some bee researchers suggest that she bites off the drone's genitalia. The fertilized queen returns to the hive and the drone dies.

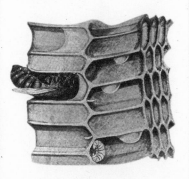

Cell structure and bee activity

THE NEW QUEEN'S DEVELOPMENT

A few days after destroying potentially rival queens (when they are still at their larval stage), the virgin queen leaves the hive and is mated by a drone. She returns to the hive to continue the colony's existence.

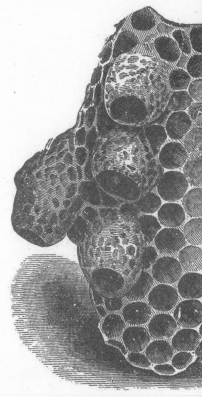

A comb from a nest of wild bees

SEARCHING FOR NECTAR

Because queen bees, drones and worker bees have their own feeding agendas within the social structure of a hive, their mouthparts differ and are specialized for their individual needs. Evolution has enabled each of them to fulfil their roles to perfection.

WHY ARE MOUTH-PARTS DIFFERENT?

Worker bees are adapted to forage and collect nectar, while the queen and drones have much smaller and shorter mouthparts; the queen relies on being fed by worker bees. Drones, however, will solicit food from returning foraging worker bees, as well as taking food stored in the hive.

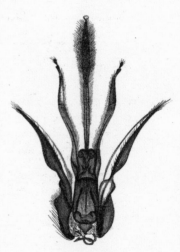

Bee mouth

LAPPING, LICKING OR SUCKING?

A worker bee's tongue is able to lap, lick and suck food; when sucking nectar from the bottom of deep flowers, it uses its proboscis like a straw. Where a flower contains little nectar, however, the bee laps it up with the tip of its proboscis, using it like a spoon. Bees are also able to lick nectar.

DETECTING TASTE

A worker bee mainly detects taste through its proboscis. Its tongue is red, long and hairy and able to curl up like a thin worm when not in use. However, when extended it is about half the bee's length. It is also thought that bees can detect and analyse tastes by using their

Collecting nectar

approximately 20% water. The removal of excess moisture occurs in the hive when a returning foraging worker bee gives most of its collected nectar to a bee within the hive. The nectar is manipulated by these 'household' bees and repeatedly exposed to the drying influence of air within a hive. This removes a large part of the moisture originally contained in the nectar.

long feelers (antennae) and legs. Research indicates that when a bee steps in a sweet syrup it will stretch out its proboscis and lick at the substance, but if a leg is put in a salt solution the bee will not give it any attention.

DRYING·NECTAR

When collected from flowers, nectar contains about 60% water, whereas honey is formed of

Where the bee sucks, there suck I:
In a cowslip's bell I lie;
There I couch when owls do cry.
On the bat's back I do fly
After summer merrily.
Merrily, merrily shall I live now
Under the blossom that hangs on the bough.

WILLIAM SHAKESPEARE (1564–1616)
The Tempest

COLLECTING POLLEN

Together with nectar and water, pollen is essential for the health and development of bees. It provides the protein essential for the development of larvae and young adult bees. Bees derive moisture from nectar, as well as collecting water from droplets on grass, and from ditches and puddles.

TRAFFICKING IN POLLEN

It is, of course, flowers that provide bees with pollen, and transporting it to the hive is a major logistic task. An estimate of the amount of pollen an average-sized colony of bees needs each year is 22–44 kg (50–100 lb), and this is claimed to represent 2–4 million bee journeys.

Bee with pollen

up of oils, fats and waxes. Pollen also contains small amounts of salts of calcium, magnesium, iron, potassium and phosphorus.

Grain of pollen

POLLEN CONTENT

The amount of protein in pollen varies widely, and in North America research suggests that it is in the range of 7–30% in native plants. About 20% (by weight) of fresh pollen is water, and 5% is made

POLLEN FORAGING

Depending on the flowers and the pollen they contain, a bee may take three or four minutes to collect a full load; at other times, this can be 20 or more minutes. If a bee is collecting both nectar and pollen at the same time, the period is even greater.

TRANSPORTING THE POLLEN

When foraging for pollen, a worker bee invariably becomes covered with pollen. She then hovers and performs the ritual of gathering, placing and compacting it in pollen sacs attached to her rear legs. This grooming ensures that no pollen is wasted, and it is all collected and taken to the hive.

ON RETURNING TO THE HIVE

When arriving back at a hive, a pollen-laden bee searches for an empty storage cell, or one that is partly filled with pollen. The bee supports herself on the edge of the storage cell, dangles her rear legs into the gap and uses her middle legs to brush off the pollen. Another bee will then push the pollen firmly into the cell. It is sometimes coated with honey; later, it is capped with wax.

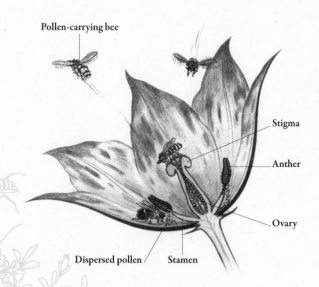

Pollen-carrying bee

Stigma

Anther

Ovary

Dispersed pollen   Stamen

Pollen collecting

THE BEE DANCE

The waggle dance is the flight pattern of bees exploring and seeking flowers that offer nectar and pollen. It also indicates to fellow bees the direction and distance of patches of flowers that yield this food, and where water and new housing locations can be found.

CONFUSED MESSAGES

Earlier it was thought that bees performed two distinct flight dances – round dances and waggle dances. Round dances were considered to indicate nearby food sources, and waggle dances to reveal distant sources. However, this is now thought to be incorrect, with a round dance basically being a waggle dance with a short waggle.

You are my honey, honeysuckle,
*   I am the bee,*
I'd like to sip the honey sweet,
From those red lips, you see,
I love you dearly, dearly,
And I want you to love me,
You are my honey, honeysuckle,
*   I am the bee.*

ALBERT H. FITZ (LYRICS)
WILLIAM H. PENN (MUSIC)
Bluebell in Fairyland, 1901

EVOLUTIONARY DEVELOPMENT

Ancestors of modern honey bees are claimed also to have given primitive messages in their flight patterns, which included zigzagging, shaking, buzzing and bumping into fellow bees. These movements can be seen in today's wasps, stingless bees and bumble bees. Evolution, however, has now given modern honey bees greater communication skills and improved foraging success.

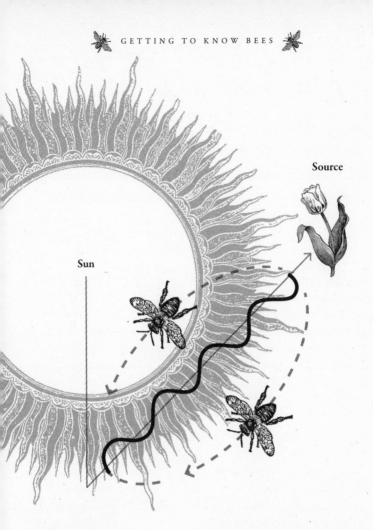

Source

Sun

COMPLEX MOVEMENTS

The waggle dance of a honey bee has a figure-of-eight pattern (as shown in the diagram above), which often appears confused because of the rapid side-to-side motions of bees. Indeed, such movement usually appears blurred.

The direction and duration of these waggle runs indicate the direction and distance of groups of flowers ready to be foraged. Flowers located directly in line with the sun are indicated by waggle runs in an upward direction: directions to the left or right are given through sideways movements.

STINGS & WARNINGS

Bees are formidable foes when enraged and adopting a defensive role. Both queen bees and workers are able to sting, but the queen usually reserves her stinging ability for killing another queen or a worker bee. Worker bees who have used their sting invariably crawl away and die.

STINGING APPARATUS

Invariably it is worker bees that attack intruders (including human beings) with their modified ovipositor. This stinging apparatus is normally hidden inside a cavity, the 'sting chamber', at the tip of a bee's abdomen, from which it can quickly protrude when required.

FORMIDABLE WEAPON

Bee stings are complex and ingenious mechanisms, used for both defence and attack. The sting itself is a straight, tapering tube with a swollen base. There are also two barbed 'lancets' which pierce the prey's skin and, with rapid alternating thrusts, sliding back and forth, become deeply embedded in the foe. These same movements activate poison pumps that drive venom into the victim's tissue.

BARBED CONSEQUENCES

Because each lancet has ten barbs, it is impossible for the bee to remove them from the victim's flesh. The motions of inserting the lancets and attempting to pull them out happens at great speed, but nevertheless leaves the bee mortally wounded and usually it soon dies. Even after the bee breaks away, the barbed lancets continue to force themselves deeper into the flesh.

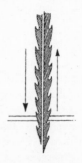

Bee-sting lancets

BEE VENOM

The amount of venom introduced into a victim is small and weighs less than half a milligram. Such is the sophistication of a bee's ability to sting that an associated gland provides a lubricant for the complex chemical composition of the venom.

OUCH!

When stung, the average human being feels pain, with the pierced part becoming red and the surrounding area pale. Most sufferers soon recover, with localized swelling and itching. However, there is a risk of severe shock and collapse – if this shows signs of happening, immediately contact a doctor or hospital.

SMOKE

Bees are terrified of fire, and its first sign is smoke. When bees detect soft, cool, gentle smoke, they rush into the storage area of a hive and gorge themselves on honey, which pacifies them and makes them more controllable when the hive is opened. However, masses of hot smoke infuriates them and causes an aggressive reaction. A smoker is a simple device in which corrugated paper or strips of sacking are burned and the smoke is puffed out gently by attached bellows.

An early bee smoker

BEE LOOK-ALIKES

Many insects resemble honey bees; some of these live solitary lives and others live in groups. They all play a role in nature, and if you sit in a garden on a summer's day many can be seen. Most are passive and will not attack you; others, such as wasps and hornets, can be aggressive and dangerous.

PARASITIC WASPS

Aggressive and with a predatory nature, these wasps catch insects which they sting into immobility and store in burrows as food for their larvae (grubs). They prey on caterpillars, weevils, flies, bees, froghoppers and spiders.

Bumble bee

BUMBLE BEES

Popular social insects, the queen forms a nest on her own (unlike queens of honey bees, hornets and stingless bees). She forms a colony of 50–200 at peak population, in about mid- and late summer.

Solitary bee

SOLITARY BEES

These form nests in hollow reeds, holes in wood or, frequently, tunnels in the ground. They are either stingless or unlikely to sting you unless threatened.

NOCTURNAL BEES

These unusual bees live in deserts, where daytime temperatures are high and prevent flight. They have greatly enlarged eye parts which are highly sensitive to light, enabling them to see better at night when foraging for food.

Nests

BEE-FLIES

Distinctive and attractive adaptations of flies, with only one pair of wings (unlike honey bees which have two pairs), they have large, fuzzy bodies, long legs and make a buzzing noise when flying. The main characteristic is a long, snout-like proboscis at the front of the head. They are usually seen in late spring and early summer.

SPIDER-MUNCHING WASPS

Mainly warm European native wasps, usually black and with red on their hind legs, they perform a dance which confuses spiders and enables the wasp to nip them with a paralysing sting. The victim is then sealed in a tunnel, with a wasp egg laid on top. On hatching, the young grub tucks into a meal of fresh food.

WASPS & HORNETS

Similar to bees, these are social insects but notorious for their ability, when disturbed and on the defensive, to sting and cause dramatic pain. They are therefore best avoided and given plenty of space. Should you have problems with them, call in a professional pest eradicator.

Yellow-jacket wasps' nest

WASPS

These are many and varied, but the one most often seen in Europe is the Common Wasp (*Vespula vulgaris*). Queen wasps overwinter in secluded places, usually holes in trees and buildings. When they emerge in spring, they are larger than queens normally seen in summer. However, they all have the distinctive patterning of yellow with black striping.

Cone-shaped nests: Usually formed of shavings and splinters of wood chewed into a yellow or grey, paper-like material. By the end of summer it may be the size of a football. During this period, the queen has been laying eggs and producing fresh broods of wasps.

Autumnal change: Towards autumn, the queen produces males (drones) and females able to

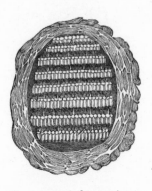

Cross-section of a wasps' nest

develop into young queens. They mate, and only fertilized females are able to survive through winter; the rest die.

Warning message in a wasp's colour

HORNETS

These are large, wasp-related social insects; some species reach 5.5 cm (nearly 2¼ in) long. There are many species of hornet, but the one most often seen in Europe is *Vespa crabro*, mainly known as the European Hornet or, occasionally, the Old World Hornet or Brown Hornet.

During spring: Fertilized queens form nests in sheltered places such as hollow tree trunks. The queen forms cells in which to lay eggs, hatching after five to eight days. These are mainly females, which become 'workers' and carry out the tasks previously undertaken by the queen. They do not lay eggs.

Hornet

In late summer: The colony grows and by late summer often reaches a peak of 700 workers. At this stage, fertilized eggs develop into females and unfertilized ones into drones (males). These mate, and the males die shortly after mating. Fertilized queens are able to survive throughout winter, but the rest of the colony dies.

PREDATORS OF BEES

Honey, wax and bees are lures to many predators, especially on sunny days when bees are fanning the hive and spreading the aroma of honey and wax. The range of predators is wide, with some only native to specific regions. Yet a few, such as the Varroa Mite, are widespread and cause devastation.

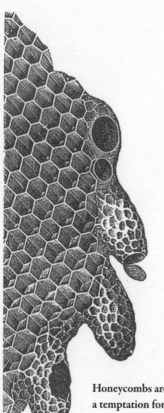

VARROA MITE

Also known as Vampire Mite and Varroa Destructor, this parasitic, pinhead-sized mite feeds on a bee's internal fluids. Unchecked, they cause death to entire colonies. Mites attach themselves to foraging bees, thereby spreading the problem. Get specialist advice if you think your bees are at risk.

MICE

These are inquisitive, especially in autumn when attempting to make a nest; they chew wax and frames. They will also urinate in the hive, deterring bees from cleaning out the hive in spring. Ensure that hive entrances are not large.

**Honeycombs are always
a temptation for predators**

Mice

BEE-EATERS

These colourful, long-beaked birds
are common in open country with
only scattered trees and bushes;
they hunt bees and other insects
in the air. They are native to
warm European areas and
North Africa. The Blue-
cheeked Bee-eater
also has a voracious
appetite for bees.

Blue Tit

OTHER BIRD PESTS

Worldwide there are many bird
pests, including shrikes, titmice,
kingbirds, finches, swifts, martins,
thrushes and mockingbirds. They
often catch bees when in flight,
although the Blue Tit in winter
taps on a hive's entrance, awaits the
appearance of a guard bee and then
makes a meal of him.

Bee-eater

MORE PREDATORS

In North America, bears are a serious pest of honey bee hives. On discovering a hive, a bear repeatedly feeds on the honey and brood chambers, eventually causing total destruction. Although frequently stung on the face, bears are not quickly deterred from obtaining honey.

Illustration for an Aesop fable

WASPS & HORNETS

These are usually after the bees, not the honey; they attack and kill foraging bees as well as those in a hive. However, in autumn they may turn their attentions to the honey. The only effective way to deter them is to eradicate their nests.

BEEWOLF

The European species *Philanthus triangulum*, a solitary and predatory wasp, is notorious for preying on bees. They sting bees, causing paralysis but not killing them. The bee is sucked dry of previously collected nectar and the body is taken to the wasp's lair for feeding larvae.

SPIDERS

Web-spinning spiders devour bees caught in their webs, while ground-hunting types may also eat them.

PRAYING MANTIS

Usually it is only by chance that a bee is caught by this intimidating insect, which grabs and devours it.

LARGE DRAGONFLIES

These soon get the better of bees, and with their strong biting parts tear their victims into pieces and eat them, often leaving only the wings.

WAX MOTHS

Adult wax moths fly at night, gaining entry to weak hives and laying eggs in the combs. Usually, these are quickly removed by the hive's bees, but if left they develop into larvae and first consume the wax combs, then honey and pollen.

CHAPTER TWO

Attracting Bees

WHAT BEES NEED

Providing bees with nectar and pollen is essential. Foraging bees also ensure the pollination of many flowering plants and consequently the healthy future of this planet. Bees require nectar as a source of energy, while pollen provides protein, which is necessary for feeding larvae. They also need water.

CREATING BEE-FRIENDLY GARDENS

Successful bee-keeping relies on nectar-rich and pollen-packed flowers. These can be native flowers as well as those in flower beds and borders; fruit trees and bushes, vegetables and herbs also play an important role.

BE 'GREEN'

Healthy bees demand plants that have not been sprayed with pesticides, fungicides or herbicides. It is best to control pests and diseases by 'green' husbandry, such as encouraging beneficial insects into your garden, regular crop rotation and 'companion planting'.

DRINKING WATER

Fresh, clean drinking water is just as essential to bees as a diet of nectar and pollen. A nearby pond, active with fish and other water creatures that prevent stagnation, is one way of providing this. Another is a fountain surrounded by stones on which bees can alight. Alternatively, a flower bed or rustic path covered with woodchips, kept wet, is a good source of moisture for bees.

COLOUR ATTRACTION

Bees are attracted to certain colours, but often not the ones we are able to see. They can see a broader spectrum of light, including ultraviolet which is invisible to us. Their specialized colour vision enables flowers to direct them to the right foraging spot for nectar and pollen.

Silverweed

Wood Anemone

COLOUR INTERPRETATIONS

The mainly white flowers of the spring-flowering Wood Anemone (*Anemone nemorosa*) appear a vibrant blue to a bee; those of Silverweed (*Potentilla anserina*), from late spring to late summer, are yellow to us but to bees have two colours, the central one indicating sources of nectar and pollen.

COLOUR CHOICES

Although bees see colours differently from us, they appear particularly attracted to flowers that to us are yellow, white, blue, purple or violet. If a gallery of plants with these colours is provided from spring to autumn, bees will be at their happiest.

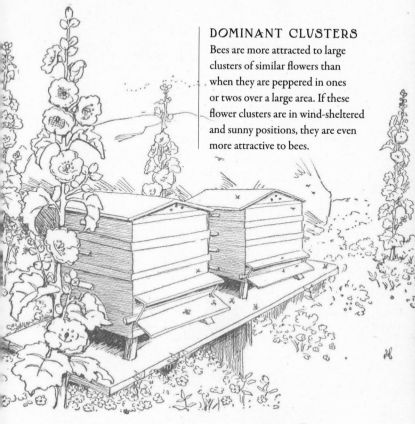

DOMINANT CLUSTERS

Bees are more attracted to large clusters of similar flowers than when they are peppered in ones or twos over a large area. If these flower clusters are in wind-sheltered and sunny positions, they are even more attractive to bees.

COLOUR-DOMINANT CENTRE

The ubiquitous Dandelion (*Taraxacum officinale*), bursting with vibrant yellow flowers sometimes throughout the year but especially during late spring and early summer, is viewed by bees as having white outer petals and dull red centres, which act as indicators for sources of nectar and pollen.

Dandelion

NATIVE OR CULTIVATED PLANTS?

Bees are usually more attracted to native plants, with their often simple flower parts, than to the highly-bred cultivated forms usually seen in flower beds and borders. The development of plants with petal-congested flowers is relatively recent, whereas bees and native plants have had a close relationship since they evolved.

DOUBLE OR SINGLE FLOWERS?

Many modern garden plants have been bred to have dominant flowers, with a large number of petals filling each bloom. These flowers have great visual appeal to garden enthusiasts, but the increased number of petals prevents bees gaining easy access to the nectar and pollen at a flower's centre. For this reason, plants with single flowers are more popular with bees than double forms.

TUBULAR FLOWERS

Plants such as foxgloves, snapdragons, penstemons and heathers have tubular flowers. These are especially popular with bees, as they are easily able to gain access to the nectar and pollen.

Snapdragon

Foxglove

Penstemon

REDUCING FORAGING DISTANCES

Allowing native plants to flourish in your garden throughout the year helps bees to reduce their foraging travels for nectar and pollen. In early and mid-spring, bees travel an average of 708 metres (775 yards) when foraging, while in late summer the distance is something like 3.86 km (2.4 miles). A few months later, in early and mid-autumn, this reduces to 1.9 km (1.2 miles).

LATE WINTER & EARLY SPRING FLOWERS

These are invaluable to bees, especially if the weather is warm and they are starting to forage. Snowdrops, crocuses and daphnes such as the Mezereon (*Daphne mezereum*) are worth planting in a warm, sunny border alongside a path or next to a wall. Remember, however, that the scarlet berries of the Mezereon are poisonous to humans.

SCENTED ATTRACTIONS

Scents emitted by flowers play an important role in stimulating bees to investigate them, especially if they were earlier learned by bees to indicate sources of food. However, other strong fragrances will also attract them and these include clover and hawthorn.

Hawthorn

SCENT & NECTAR

It is thought that many flowers emit their strongest scents when they contain most nectar, thereby assuring the greatest chance of pollination. It is also said that bees will remember a perfume more readily than a colour. Indeed, it is further claimed that a foraging bee is more attracted by a scent than by a colour, preferring the correct perfume and wrong colour to the correct colour and the wrong perfume.

SCENTED LAWNS & PATHS

Chamomile lawns, with their deeply divided leaves which when bruised have the fragrant fusion of bananas and apples, have been known to attract bees. However, thyme-clad paths formed of Wild Thyme (*Thymus serpyllum*), also known as English Thyme, are equally attractive. Thyme-fragrant flowers

Chamomile

Thyme

range from deep red through pink to white, from mid- to late summer. To prevent thyme becoming unnecessarily trodden upon and squashed, put stepping stones down the path's centre.

❧

RESERVED LAWN MOWING

Where lawns have reverted to their natural flora, with perhaps red and white clover appearing through and above the grass, do not mow them short and frequently, as they are a rich source of food for bees, especially in autumn. Both of these clovers are valuable to bee-keepers, particularly the prostrate white form that is often known as Kentish Clover.

Clover

SCENTED WILD PLANTS

The range of countryside plants that produce bee-attracting fragrances is wide, and many are popular in literature. A walk along a country hedgerow during summer highlights many rustic plants that entice bees, including the Sweetbriar, Traveller's Joy and Wayfaring Tree, all richly scented.

Whiles yet the dew's on ground,
gather those flowers,
Make haste. Who has a note
of them?

WILLIAM SHAKESPEARE
(1564–1616)
Cymbeline

RICHLY REDOLENT

Whereas most plants visited by flies have an unpleasant smell, those that attract bees have sweet and fruity redolences. Fragrances given off by plants are intensified during warm weather, with bees visiting flowers of the same species for as long as possible before moving on to another type.

BEE-ENTICING PLANTS

There are many attractive wild plants to choose from, including the following:

🐝 **Borage** (*Borago officinalis*): loose clusters of bright blue flowers with purple stamens.

🐝 **Cowslip** (*Primula veris*): drooping clusters of yellow flowers with orange spots at their centres.

🐝 **Evening Primrose** (*Oenothera biennis*): sometimes known as the Common Evening Primrose, with masses of pale yellow flowers.

🐝 **Lily of the Valley** (*Convallaria majalis*): arching stems bearing five to eight white and waxy bell-shaped flowers.

🐝 **Musk Thistle** (*Carduus nutans*): thistle-like bright red-purple, nodding flowerheads.

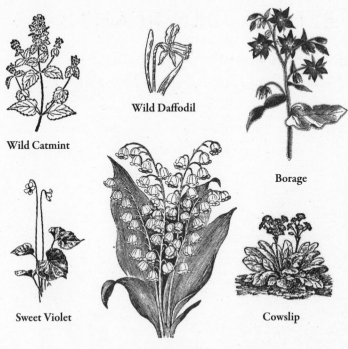

Wild Daffodil

Wild Catmint

Borage

Sweet Violet

Cowslip

Lily of the Valley

🐝 **Sweet Violet** (*Viola odorata*): heart-shaped leaves and blue-violet or white flowers.

🐝 **Wild Catmint** (*Nepeta cataria*): tooth-edged, heart-shaped and mint-scented leaves, and red-dotted white flowers.

🐝 **Wild Daffodil** (*Narcissus pseudonarcissus*): also known as the Lent Lily, with trumpet-shaped, slightly nodding, yellow flowers.

*Soon will the high Midsummer
    pomps come on,
Soon will the musk Carnations
    break and swell,
Soon will we have the gold-
    dusted Snapdragon,
Sweet William with his homely
    cottage-smell,
And stocks in fragrant blow.*

MATTHEW ARNOLD (1822–1888)
Thyrsis

BEE-GARDEN PLANTS

Many flowering plants that burst into colour in beds and borders in our gardens have been developed by seed companies from 'wild' plants, either those native to this country or from warmer climes. Often, their range of colours has been enhanced and flowering times extended.

Bring hither the pink and purple columbine,
*   With gillyflowers:*
Bring coronations, and sops-in-wine,
*   Worn of paramours.*

EDMUND SPENSER (1552–1599)
The Shepherd's Calendar

WIDE RANGE OF GARDEN PLANTS

There are many bee-enticing garden flowers, including hardy annuals (sown where they are to grow and flower), half-hardy annuals (raised in gentle warmth in spring and planted when all risk of frost has passed), herbaceous perennials (these live from one season to another, dying down to soil level in autumn and reappearing the following spring), and trees and shrubs. Many of these are described on this and the following pages.

HARDY ANNUALS

❧ **Annual Tickseed** (*Coreopsis tinctoria*): many varieties, in shades of crimson, golden-yellow or maroon.

❧ **Cornflower** or **Field Cornflower** (*Centaurea cyanus*): sprays of pink, red, purple, blue or white flowers.

❧ **Mignonette** (*Reseda odorata*): cottage-garden flower, with richly scented, small, yellow-and-white flowers in loose heads.

❧ **Poached Egg Plant** (*Limnanthes douglasii*): spectacular, with yellow-centred, white flowers.

Annual Tickseed

❧ **Sweet Scabious** (*Scabiosa atropurpurea*): clustered heads on long stems. Wide colour range – dark purple, red, pink, lavender, salmon and red.

❧ **Sweet Sultan** (*Centaurea moschata*): thin, stiff stems bearing white, yellow, pink or purple flowers.

SEASONAL FLOWERS

Half-hardy annuals are raised in gentle warmth in spring and planted into beds and borders when all risk of frost has passed. Many of them are native to warm climates where they naturally grow as hardy annuals. Raising plants early in the year extends their flowering periods.

NONCONFORMIST PLANTS

Some plants do not always obey our strict growth classifications. For example, in cool climates antirrhinums can be sown outdoors where they are to flower, or raised slightly earlier in greenhouses for later planting into borders and beds. Yet in warm regions they establish themselves as hardy perennials and survive the year without any protection.

The odour of plants and the
 fresh sighte,
Wolde have made any heart
 lighte,
That ever was born.

GEOFFREY CHAUCER
(1340–1400)
The Franklin's Tale

❀ **Cherry Pie** (*Heliotropium arborescens*): half-hardy perennial usually raised as a half-hardy annual, with tightly packed heads of flowers that range in colour from dark violet through lavender to white.

❀ **Mexican Aster** (*Cosmos bipinnatus*): Half-hardy annual with varieties in colours including white, rose-pink, crimson, orange and yellow.

❀ **Snapdragon** (*Antirrhinum majus*): can be raised as a half-hardy annual, hardy annual or, in some areas, can become a hardy perennial. Distinctive dragon-like heads in many colours.

❀ **Sweet Alyssum** (*Lobularia maritima*): sometimes known as Sweet Alison, this hardy annual is often grown as a half-hardy annual, with flower clusters in white or various shades of purple.

❀ **Vervain** (*Verbena* × *hybrida*): half-hardy perennial usually raised as a half-hardy annual, with clustered heads of brightly coloured flowers in a range including scarlet, blue, carmine and white.

Snapdragon

Sweet Alyssum

Vervain

HERBACEOUS PERENNIALS

These garden plants die down to soil level in autumn, remain dormant underground throughout winter, and produce fresh shoots in spring. They live in borders and beds for several years before becoming congested; then they need to be dug up and divided, and the young parts replanted.

ATTRACTING BEES

There are many beautiful bee-enticing herbaceous plants to consider, including:

🐝 **Astilbe** (*Astilbe × arendsii*): fern-like leaves and loose, fluffy, pyramidal clusters of flowers; several colours, including white, red and pink.

🐝 **Bergamot** (*Monarda didyma*): also known as Bee Balm and Oswego Tea, with densely whorled heads of flowers in colours from pink to scarlet and white.

🐝 **Bugle** (*Ajuga repens*): whorls of blue, tubular flowers with protruding lips on erect shoots.

Bugle

🐝 **Foxglove** (*Digitalis purpurea*): usually raised as a hardy biennial, but often perennial, with tall, upright stems bearing tubular, purple, pink, red or maroon, spotted flowers.

🐝 **Globe Thistle** (*Echinops ritro*): globular, steel-blue flowers.

❧ **Hellebore** (*Helleborus argutifolius*): evergreen, three-lobed leaves and cup-shaped, yellow-green flowers.

❧ **Jerusalem Cowslip** (*Pulmonaria officinalis*): white-spotted green leaves and funnel-shaped, purple-blue flowers.

❧ **Lamb's Ears** (*Stachys byzantina*): mid-green leaves densely covered with white, silvery hairs which create a woolly appearance. Spikes of purple flowers.

❧ **Mullein** (*Verbascum phoeniceum*): graceful spires of white, pink, purple, mauve or blue flowers.

❧ **Pincushion Flower** (*Scabiosa caucasica*): almost leafless stems bearing lavender-blue flowers. Varieties with white, violet-blue and light-blue flowers.

Hellebore

❧ **Thrift** (*Armeria maritima*): hummock-forming, with pink flowerheads.

❧ **Vervain** (*Verbena bonariensis*): tall, branched stems bearing clusters of primrose-shaped, rose-lavender flowers.

❧ **Yellow Loosestrife** (*Lysimachia punctata*): long-lived, with upright stems densely covered in whorls of bright yellow, cup-shaped flowers.

Thrift

*Borage and hellebore fill
  two scenes,
Sovereign plants to purge
  the veins,
Of melancholy, and cheer
  the heart,
Of those black fumes which
  make it smart.*

ROBERT BURTON (1577–1640)
Anatomy of Melancholy

47

TREES & SHRUBS

Trees and shrubs are, until the inevitable arrival of their old age and need of replacement, permanent features in gardens. Some of them, such as the Common Privet, Holly and heathers, can be found in the wild but nevertheless are worthy of places in bee gardens.

✤ **Common Privet** (*Ligustrum vulgare*): deciduous or semi-evergreen shrub with lance-shaped leaves and clusters of white flowers.

Guelder Rose

✤ **Guelder Rose** (*Viburnum opulus*): deciduous shrub with maple-like, dark green leaves and flat heads of white flowers.

✤ **Heather** (*Calluna vulgaris*): evergreen shrub with scale-like, stubby leaves and terminal clusters of single or double flowers, in shades of pink, purple or white.

✤ **Holly** (*Ilex aquifolium*): evergreen, shrub or tree, with glossy-green, spine-edged leaves and dense clusters of white flowers.

Old English Lavender
(*Lavandula officinalis*): evergreen
shrub with narrow, silver-grey
leaves and spikes of pale grey-blue
flowers. Several varieties, in colours
including deep purple-blue.

Rosemary (*Rosmarinus
officinalis*): evergreen shrub with
aromatic, dark green leaves and
clusters of mauve flowers.

Thyme (*Thymus* spp.): a wide
range of thymes lure bees to their
flowers, including the Wild Thyme
(*Thymus serpyllum*) with rounded
clusters of flowers, ranging in
colour from deep red through
pink to white.

Lavender   **Rosemary**

With marjoram knots,
    sweetbrier and ribbon-grass,
And lavender, the choice of
    every lass,
And sprigs of lad's-love, and
    familiar names,
Which every garden through the
    village claims.

JOHN CLARE (1793–1864)
The Shepherd's Calendar

BEES AS POLLINATORS

Pollinating insects are essential for the continuation of life on this planet, and without their skills the growth and development of food crops would diminish and eventually fail. Bees are the main pollinators of food crops, and need all the help and encouragement we can give them. The fruits, vegetables and nuts indicated on these pages are food plants that rely on bees for pollination and their future growth.

FRUITS

- Apple
- Apricot
- Avocado
- Blackberry
- Blackcurrant
- Blueberry
- Boysenberry
- Cantaloupe Melon
- Cranberry
- Greengage
- Guava
- Kiwifruit
- Lemon
- Lime
- Loquat
- Lychee
- Mango
- Nectarine
- Peach
- Pear
- Persimmon
- Plum
- Pomegranate
- Quince
- Raspberry
- Redcurrant
- Sour Cherry
- Strawberry
- Sweet Cherry
- Tangerine
- Watermelon

Blackberry

VEGETABLES
- Aubergine (Eggplant)
- Beetroot
- Broad bean
- Broccoli
- Brussels Sprouts
- Cabbage
- Carrot
- Cauliflower
- Celery
- Chilli Pepper
- Chinese Cabbage
- French bean
- Gourd
- Haricot bean
- Kidney bean
- Lima Bean
- Marrow
- Onion
- Pumpkin
- Runner bean
- Soybean
- Squash
- Turnip

NUTS
- Cashew
- Chestnut
- Coconut

Squash

CHAPTER THREE

Where do Bees Live?

HONEY BEES IN THE WILD

The honey bee is a social insect, and for thousands of years before being domesticated they lived wild in groups. The evolution of bees was mirrored by the development of flowering plants and their need to be pollinated, ensuring the future for both plants and bees.

Part of a drawing discovered in 1919 on a cave wall in Spain that shows a person gathering wild honey from a cliff cavity

WARM & WIND-PROTECTED

Bees in the wild have a clear preference for their nesting positions. In the northern hemisphere, they seek sites in south-facing positions sheltered from the wind. In the southern hemisphere, north-facing areas are preferred.

A cluster of swarming bees searching for a place to rest and form a new home

NESTING CAVITIES

Not all nesting cavities suit honey bees, and they prefer spaces about 45 litres (1 cubic foot) in volume; they avoid areas smaller than 10 litres (⅓ cubic foot) and larger than 100 litres (3 cubic feet). They also have a preference for the height of a nesting area, selecting those between 1 m (3 ft) and 5 m (16 ft) above the ground.

WILD NEST CONFIGURATIONS

Nests of wild bees are usually formed of multiple honeycombs, hanging downwards and parallel to each other, with uniform spaces between them. The nesting area usually has just one entrance. The bees usually remain in their nest for several years and until they are congested and need, again, to swarm.

COLONY EXPANSION

Where bees have swarmed, they usually create a new home at a distance of about 300 m (980 ft) from the parent colony.

Wild bees forming honeycomb nests

EARLY HIVES

Before being domesticated, honey bees took advantage of holes in trees, recesses in walls and discarded pots. Then they thrived in early man-made hives or 'skeps', which protected them from wet weather and provided insulation against extremes of cold and warmth throughout the year.

ABYSSINIAN SEWN HIVES

These hives were made from rolled bark wrapped in straw before being tied and sewn. There were also hives formed by hollowing trunks of kolkual cactus trees.

WILLOW-TWIG SKEPS

Woven from flexible willow stems, with a detachable top, these were firm and durable, with an angled, stable base.

OLD FRENCH SKEPS

Early French skeps were formed of seasoned straw vertically secured to a simple frame and with a rain-proof, knotted and twisted top. A small, single, ramped entry at the base of the skep enabled bees to enter and leave. These early skeps were widely known in northern and western Europe, and later replaced by coiled skeps.

COILED STRAW SKEPS

Coiled skeps took many forms, including later ones with removable tops. A person who made coiled and woven skeps was known as a 'skepper'. This surname, in certain forms, still exists today.

HOLLOWED TRUNKS

Dried trunks from small trees, as well as large cacti, were hollowed and raised on stout, forked supports to keep them dry and free from vermin. Sometimes, Abyssinian sewn hives were supported horizontally.

Tanging (hitting metal pots and pans) was said to encourage a swarm of bees to settle

EARLY BEE-KEEPING

The above illustration from Pietro Andrea Mattioli's *Commentaires*, first published in Italian in 1544, shows early hollowed-log hives positioned vertically and in groups. Mattioli was a doctor and naturalist. Apart from his botanical and naturalist investigations, he is claimed to be the first doctor to describe allergies to cats, from which he suffered.

SKEPS

Few pieces of bee-keeping equipment have a more bucolic appearance than a skep (usually formed of coils of dried grass or straw). They were widely used throughout Europe and provided homes in which bees could live and from which honey could be harvested.

DESTRUCTIVE IMPLICATIONS

Skeps are simple and inexpensive to make, but have several disadvantages: the well-being and development of the bees inside the skep could not be easily monitored, and the removal of the honey resulted in the total destruction of the colony. Bee-keepers either killed the bees or used smoke to scare and drive them out of the skep.

SKEP DEVELOPMENT

To extract the honey, skeps were often squeezed in a vice (fortunately this is now illegal in many countries). Later developments for skeps included a small, woven basket at the top of the main skep in which the bees could store honey (this is the equivalent of today's 'super' – see page 67). A small hole enabled bees to pass from the lower part (which nowadays would be called a 'brood chamber') to the upper area.

BEE SHED

Where several skeps were kept they were often put in a bee shed to help repel vermin (including birds), keep the skeps dry and longer-lasting, and to make 'bee-husbandry' activities easier. However, few additions to bee-keeping would have been more useful than to have protective clothing.

CREATING EXTRA SPACE

A development in Scotland to enable extra space to be provided in the brood chamber for the rearing and development of bees was known as an 'eke'; using this became known as 'eking'.

Coiled straw forming three distinct, movable sections to a hive

FENDING OFF VERMIN

To keep mice and other vermin from entering a skep, it could be raised on a small, round plinth that included a landing area for the bees.

OBSERVATION HIVES

The English diarist Samuel Pepys (1633–1703) is claimed to have had an early observation hive, and from the middle of the 1800s they were widely fashionable, especially among the gentry. For this clientele, beehive makers constructed highly ornate observation hives.

Samuel Pepys

NOVEL PURSUIT

Apart from their novelty value, observation hives offer the opportunity to study bees close-up throughout the year, and without the risk of being stung. At the same time, the bees suffer minimal disturbance. Observation hives are widely popular exhibits at craft fairs and for use in educational establishments.

A few early observation hives were complex and formed of several frames, while others were simple and had only one frame

WHAT ARE THEY?

Modern observation hives are usually only one frame in depth and two frames high (some are just one frame high). The construction is usually of wood, with sides formed of glass, and with provision for feeding and ventilation. An entrance tube is necessary, linking the hive with the outdoors. Hinged or clip-in side panels are needed to provide darkness and privacy for the bees when not being studied.

EARLY DESIGNS

Their designs are varied and ingenious, with many dating back to the 18th century. Some are connected through a tube to the outside, while others are permanently outdoors.

18TH-CENTURY DESIGN

An early observation hive from the 18th century is shown on the right. An entrance for the bees can be seen low down on the hive's side.

RAUMUR'S HIVE

Ren Antoine Ferchault de Raumur (1683–1757), a French scientist, contributed to many scientific studies, including those of insects. Here is one of his observation hives when studying bee biology.

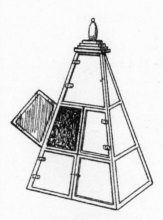

Raumur's hive

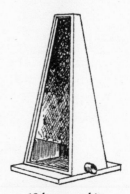

18th-century hive

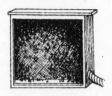

GLASS HIVE

An early glass observation hive with a reinforced flexible tube between the hive and a hole in a window frame.

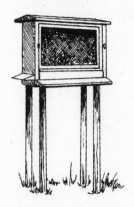

GARDEN HIVE

A simple observation hive for positioning in a garden. The stout supports keep the hive free from vermin and enable the bees to be easily inspected.

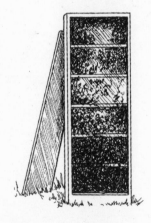

SIX-FRAME HIVE

An observation hive with six frames.

LARGE BROOD FRAME

A later development, with a large 'brood' frame near its base and a 'super' formed of 6–8 frames.

MODERN OBSERVATION HIVE

This design from the 1950s is formed of two deep frames, a wooden framework and glass sides. It is ideal for positioning on a windowsill, with the entrance extending though a hole bored in the window's frame or its surround.

OUTDOOR SCIENTIFIC OBSERVATION HIVE

In earlier times, the Soviet Union undertook detailed research into the flying activities of bees. This six-frame outdoor observation hive (see below) enabled bees to be studied in as a natural a setting as possible. To reduce the chance of an observer being confused by large numbers of bees entering and leaving at the same time, the hive is fitted with a long, glass 'lobby'.

This enabled the movements of bees to be tracked when they returned from foraging and to discover how pollen and nectar were treated when taken into a hive. Some worker bees had the sole task of receiving pollen and nectar from returning bees and played a fundamental role in the hive's activities.

Scientific outdoor observation hive

MODERN HIVES

Hives we would recognize today first appeared in the 19th century, although they owe much to hive development in the 18th century. The search for better and more practical hives was pursued in many countries, including Russia, Germany, France and Britain, as well as North America.

EARLY MODERN DEVELOPMENTS

The transition from skeps – where the entire colony was destroyed in order to harvest the honey – was first described in about 1770. This innovation involved positioning a further skep above the existing one and fitting wooden bars in it to which bees could secure their combs. Access for bees was through a small hole made in the top of the base skep.

SEPARATE CHAMBERS

The technique of separating the brood area (where eggs are laid and develop into adult bees) from the super (where honey is stored) was pivotal in the development of modern hives. It enabled frames to be easily inspected.

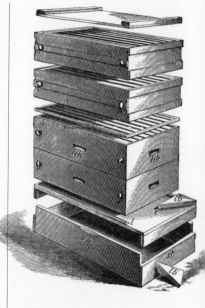

Heddon's hive – James Heddon, born in 1845 in North America, had more than 500 colonies of bees in 1879; he went on to greater fame by making the world's first artificial fishing lure

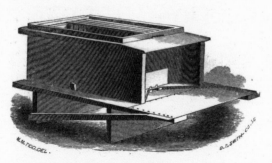

An early Langstroth hive

LANGSTROTH HIVES

In the early 1850s the Reverend Lorenzo Lorraine Langstroth (1810–1895), a native of Philadelphia, discovered that by leaving precise 'bee-spaces' around frames the bees did not fill them with wax. This enabled frames to be removed without significantly disturbing the bees.

EARLY BEE BOOK

In 1853, the Rev. Langstroth published *The Hive and the Honey Bee*, in which he described the fundamental dimensions of the modern beehive. The book is a classic in bee literature and has been reprinted many times. He is often said to be 'the father of modern bee-keeping'.

NATIONAL HIVES

This design is widely popular in the United Kingdom, with a square, somewhat box-like outline. The frames are smaller than in the Langstroth design, and some apiarists suggest that the brood chamber is too small for the increased egg-laying ability of modern strains of queen bees. Therefore many bee-keepers use the normal brood chamber together with a super to accommodate this greater vigour.

National hive

COUNTRY-COTTAGE HIVES

The late 19th-century English cottage garden, with its relaxed and informal arrangement of flowers, herbs, vegetables and fruit trees, inevitably housed a hive of bees that provided honey, an essential food for the health and well-being of families.

THE WBC HIVE

Invented by William Broughton Carr (1837–1909) who was born in Yorkshire, the WBC hive has a double-wall construction which keeps the bees cool in summer and warm in winter. The design was published in the 1890s and its construction resembles a National hive with a surround formed of removable sides.

It has a classic 'country-garden' style, often seen in pictures and paintings. However, it is not often used by commercial bee-keepers as the need to remove the sides every time attention is given to the hive becomes a time-wasting task.

WBC hive

I *Honey-flowers to the honey-comb,*
   And the honey-bee's from home.

   A honey-comb and a honey-flower,
   And the bee shall have his hour.

   A honeyed heart for the honey-comb,
   And the humming bee flies home.

   A heavy heart in the honey-flower,
   And the bee has had his hour.

II *A honey-cell's in the honeysuckle,*
    And the honey-bee knows it well.

    The honey-comb has a heart of honey,
    and the humming bee's so bonny.

    The honey-flower's the honeysuckle,
    And the bee's in the honey-bell.

    The honeysuckle is sucked of honey,
    And the bee is heavy and bonny.

DANTE GABRIEL ROSSETTI
(1828–1882)
Chimes (excerpt)

CHAPTER FOUR

Bee-Keeping Equipment

HIVES FOR TODAY'S BEES

Modern bee-keepers have several types of hives available to them. Some of these are further developments from the early Langstroth hives and include the popular National type. Alternatively, many apiarists prefer the picturesque cottage-garden design of the WBC hive.

WBC HIVE

This has similar internal parts to the National hive, but with the benefit of double-wall construction. Because of its construction, the outside can be painted, usually in white.

TOP BAR HIVE

This innovative and relatively new hive design is used extensively in Africa and the Caribbean, as well as parts of Europe and North and South America. It is available ready assembled or in flat-pack form. Noted for its simplicity and eco-friendly design, it holds 24 simple bar frames and measures about 88 cm (34 in) long, 48 cm (18 in) wide and 30 cm (12 in) high. For many apiarists it is an ecomonic way to start keeping bees.

NATIONAL HIVE

This single-wall hive design is widely used commercially: it enables the various parts to be easily and confidently handled when opening a hive and attending to the bees. A stand raises the hive about 25 cm (10 in) above the ground, clear of weeds and reducing the risk of predators attempting to gain entry. When painting, use only a recommended paint or linseed oil. Household paint usually seals the wood's surface and causes internal condensation that encourages the development of moulds.

Roof
This hive's roof has a covering of galvanized metal; types with gabled roofs are also available. Ventilation slots in the roof are essential.

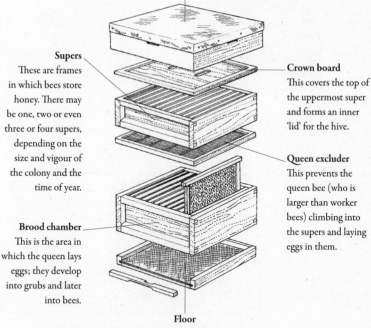

Supers
These are frames in which bees store honey. There may be one, two or even three or four supers, depending on the size and vigour of the colony and the time of year.

Crown board
This covers the top of the uppermost super and forms an inner 'lid' for the hive.

Queen excluder
This prevents the queen bee (who is larger than worker bees) climbing into the supers and laying eggs in them.

Brood chamber
This is the area in which the queen lays eggs; they develop into grubs and later into bees.

Floor
Forms the base of a hive and provides an entrance for bees. It also usually includes a Varroa Mite mesh to give protection from these pernicious pests.

National hive

INSIDE HIVES

The internal parts of a hive are tailored to confine the queen bee to the brood chamber. However, worker bees are able to roam freely within the hive, organizing the laying of eggs, storing food, keeping the hive clean and fanning at the entrance in summer to keep it cool.

Gravenhorst hive: combination of a skep and a framed hive

Radouan Eke hive: an early development of a framed hive

August von Berlepsch developed an open-sided hive, as well as rectangular, movable frames

SUPERS

Fitted with frames suspended at their top and outer edges, which hang within the super. Honey is stored in the frames.

Super

BROOD CHAMBERS

Similar to supers but deeper. They are fitted with frames into which the queen can lay eggs.

QUEEN EXCLUDER

Made of wire, slotted steel or plastic, it prevents the queen laying eggs in the wrong place.

FRAMES

Initially, these are flattish pieces of wax with a honeycomb structure and known as 'beeswax foundation'. They are held in a wood or plastic frame and usually kept in place with diagonal wires. Worker bees work on these frames to produce cells into which eggs can be laid or ones for storing food.

Frame

PROTECTION

To ensure that keeping and handling bees is a pleasurable experience, protective clothing is essential. Illustrations of early bee-keepers opening hives and handling frames with bees covering their arms and hands may appear 'macho', but is dangerous. Never take any chances of being stung and always put on protective clothing.

EXTRACTING &
STORING HONEY

Early bee-keepers who kept bees in skeps had to destroy the colony to be able to collect the honey. Nowadays, with the introduction of hives with removable frames, it is possible to extract honey yet leave the hive intact and the colony of bees healthy and unharmed.

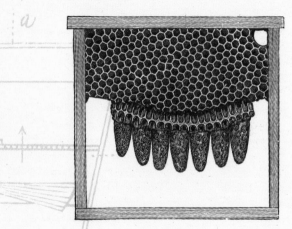

Frame and honey

MECHANICAL HONEY EXTRACTORS

These are mechanical devices, powered either manually or by electricity, which rotate frames full of honey within a stainless-steel or polythene barrel-type container where the honey can be collected. Care is needed to prevent the frames being damaged.

Modern extractor

German design   Cowan extractor

EARLY DEVELOPMENTS

Early honey extractors appeared in the mid-1860s, when the American apiarist A. J. Root designed a centrifugal honey extractor. Other apiarists had similar thoughts, including the American Willard C. Weeks and the Austrian Major Franz Major von Hruschka.

ALTERNATIVE METHODS

Another way to extract honey from frames is to use a hot, sharp knife to cut the entire honeycomb out of the frame; it can then be used as cut-comb honey. A further way is to squeeze out the honey.

STORING HONEY

Because it can be easily stored, honey has long been a favourite food. Put it in tightly covered containers, and keep it relatively cool, 10–18°C (50–64°F), away from direct and strong sunlight. Do not put it in a refrigerator.

CLOUDY HONEY

Often, stored honey becomes cloudy (known as crystallization); this is not an indication of deterioration. Honey which has crystallized can be re-liquefied by placing the jar of honey in a pan of hot water until it clears.

CHAPTER FIVE

Talking to Bees

INVOLVING THE BEES

In many parts of Europe, as well as in North America, it was customary in earlier years to tell bees about changes within households, such as deaths and weddings. Bees and their honey were so precious to families that they did not wish to upset them and cause them to die or go away.

BEE CONVERSATIONS

Most modern apiarists give little credence to the notion of talking to bees, but as far back as the 1600s poetry embraced the close relationship between bees and people. Later, the North American poet John Greenleaf Whittier (1807-1892) wrote the bucolic poem 'Telling the Bees'.

GOOD-LUCK MASCOT

A charm or brooch shaped like a honey bee is claimed to bring luck in business and thought to encourage thrift and perseverance.

FORETELLERS OF SUCCESS

Bees are famed for encouraging success and therefore should not be shooed away when flying into your house, even though they may annoy you!

EXPECT A STRANGER

Bumble bees, close relatives of the honey bee, when entering through a door, are said to foretell the coming of a stranger.

MOVING HIVES

In earlier centuries there was a strongly held belief that before moving a hive of bees to a new place it was essential to tap on its top three times and tell them of the forthcoming move. If not done, the bee-keeper could be badly stung.

WEATHER RHYMES

Accurately predicting the weather is a desire deeply embedded in most people – every evening there is another opportunity to have one's predictions come true! Invariably, bees were part of this daily prophesying, with sayings such as:

If bees stay at home,
Rain will come soon,
If they fly away,
Fine will be the day.

CLAIMING A SWARM

Swarms of bees are valuable, and in centuries past the owner of the hive from which the bees swarmed would chase them, waving and tapping a warming pan, frying pan or kettle, in an attempt to claim them for his own. This was known as 'tanging' and produced a loud ringing or clanging noise.

VALUING A SWARM

In earlier years, when honey was the main source of sweetness, bees were especially prized and protected. The earlier in the season a swarm appeared, the greater was its value. This led to the saying: *'A swarm of bees in May is worth a load of hay. A swarm of bees in June is worth a silver spoon.'*

PREVENTING SWARMS

Bees especially swarm during early summer, when the weather is warm. One way to calm them is to spray the area around the hive with water – not soaking the hive but slightly cooling it.

Traditionally, aromatic herbs such as Bee Balm (*Monarda didyma*), thyme or lavender were rubbed on hives to calm and settle bees to prevent them swarming.

Instrumento per le Api

The bees pillage the flowers here and there, but they make honey of them which is all their own; it is no longer thyme or marolaine.

MICHEL DE MONTAIGNE
(1535-1592)

74

OWNING BEES

In previous centuries, apart from bartering with corn or a small pig, gold coins were usually offered to buy a swarm. This prompted the saying: *'If you would wish your bees to thrive, Gold must be paid for every hive; For when they're bought with other money, There will be neither swarm nor honey.'*

GIFT OF A SWARM

This was generally considered to be lucky. Bartering and buying, however, were the normal ways in which bees were obtained.

SWARM OMEN

If bees swarmed onto a hedge or dead tree, this was considered to be an omen for death in a family.

LOVE & MARRIAGE

These were perennial themes in country life during earlier times, and before getting married a daughter was expected to tell the bees of her forthcoming marriage. Also, a piece of bridal cake had to be left in front of the hive, with white ribbons tied around it.

A WEDDING-DAY SWARM

If bees swarmed on a wedding day, this was taken to be an indication of forthcoming fertility, wealth and happiness.

LOVING WORDS

Honey has long been associated with love and as a term of endearment, with variations including 'Honey Bun', 'Hon' or even 'Honey-bun Lamb'. These endearments of love, invariably offered by men to women, are ways to pledge love and fidelity.

*But love is blind, and lovers
  cannot see,
The pretty follies that themselves
  commit.*

WILLIAM SHAKESPEARE
(1564–1616)
The Merchant of Venice

*The rose is red, the violet blue,
Honey is sweeter and so are you.*

ANONYMOUS

VIRGINAL CONFIDENCE

It used to be widely thought that a virgin could walk through a swarm of bees without any risk of being stung.

❧

HAPPY LIFE OMEN

In some countries, a bee flying around a sleeping child was said to indicate that he or she would have a happy life.

BEES IN PARADISE

In previous years it was thought that bees originated in Paradise, where they were known as *The Little Servants of God*. For that reason it was said to be unlucky to kill one.

❧

Love conquers all things: let us too surrender to love.'

VIRGIL (70–19 BC)

BEE SUPERSTITIONS

For centuries there was a superstition that all the bees in a hive would die if they were not told of a death in a family. In some areas, it was the custom to place funeral cake and a glass of wine near the hive. Alternatively, hives were moved to another place after a family death.

TELLING BEES OF A DEATH

It was thought to be vital for a relative to visit the beehive and to repeat three times the saying: 'Little brownies, little brownies, your master (or mistress) is dead.' In some rural areas, there was a custom of turning a hive completely around; at the same time, the bees were told of their new ownership.

PASSING AROUND A HIVE

In some areas there was a custom of carrying the deceased, when leaving home and before going to church for the committal, around the family's hive.

Bee sting

BEE STINGS

Putting mud or manure on a bee sting was claimed to decrease the pain. However, although it has a cooling quality, it encourages the risk of tetanus – which is much worse than suffering the sting!

EASING RHEUMATISM

It is widely supposed that a sting from a bee is a cure for rheumatism. But this begs the question: why do many apiarists suffer from this painful and debilitating problem?

HONEYED CONFECTION BOXES

During the 15th and 16th centuries, honey was an essential part of medicines. Together with saffron, it was used to form sugary pastilles of spices, herbs and seeds, which were frequently kept in a confection box.

O death, where is thy sting?
Corinthians

CHAPTER SIX

Honey, Mead & Wax

MAGICAL HONEY

Honey has for centuries been considered a natural elixir, especially for country folk. Apart from its medicinal qualities, it was the main sweetener and preserver of food. For this reason, bees were treated with respect and regarded as an essential part of family life.

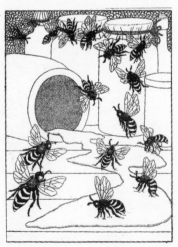

Both bees and wasps are attracted to spilled honey

Gathering honey, ancient Egypt

FLAVOURING ALE

In medieval England, ale (made from malted barley or oats, without the inclusion of hops) that turned sour was revitalized and flavoured by the addition of honey, herbs and spices.

MEDICINAL VALUE

Honey has long been known for its healing qualities. People suffering from the ague (a recurrent shivering chill or fever) were treated with a pint of warm beer and honey. External abrasions and internal ulcers were also treated with honey.

RELIGIOUS INVOLVEMENT

In Buddhism, honey is central in the festival of Madhu Purnima, commemorating when the Buddha made peace among his disciples by retreating into the wilderness. A legend suggests that a monkey and elephant fed him fruit and honeycomb during this time.

BEAUTY TREATMENT

Honey is used in many beauty treatments, including bubble baths, hand lotions, soaps and moisturizers. It attracts and retains moisture as well as having anti-microbial properties. Many anti-ageing preparations contain honey.

MEAD: THE DRINK OF INSPIRATION

Also known as honey wine, the earliest archaeological record of mead dates back to around 9,000 years ago, when it was known throughout Europe and Asia. Much has been written about it, and Pliny the Elder (AD 23–79) called it *militites* (drinkable honey) in his *Naturalis Historia*.

Pliny the Elder

FERMENTATION

Mead is created by the fermentation of honey in a mixture of water and yeast. Historically, mead was fermented by wild yeasts and the bacteria present on skins of fruit, as well as by bacteria naturally present within the honey itself.

APHRODISIAC

Mead has a long reputation of encouraging amorous endeavours, and it is claimed that the word 'honeymoon' derives from the ancient Scandinavian custom of giving newly-weds mead to drink for a month (a full moon) after a wedding to ensure the birth of a boy child.

Beowulf

MONASTIC MEAD

During medieval times, monasteries retained the tradition of keeping bees and producing mead, especially in northern climates where grapevines could not be grown with any certainty of producing crops of grapes.

DANISH INFLUENCE

In the Old English heroic poem 'Beowulf', written by an anonymous Anglo-Saxon poet and dated between the 8th and 11th centuries, Danish warriors are said to have drunk mead. Indeed, in Old English the name Beowulf literally means 'beewolf'.

MAY DAY DRINK

In Finland, Sima (a form of sweet mead) is still popular as a drink during May Day festivities. Often spices are added during secondary fermentation to temper the amount of sugars in the drink - they rise to the surface of the wine when the brew is ready to be sampled.

THE IMPORTANCE OF WAX

Beeswax has been an essential part of human activities for many centuries. It was a fundamental ingredient in wax seals to assure the authenticity of legal documents, candles and furniture polishes, and was used as a sealant and lubricant for early firearms using black powder (a mixture of sulphur, saltpetre and charcoal) as an explosive charge.

The bodies of worker bees produce a
wax which is then chewed to refine it

WAX RECIPES

In the Middle Ages, sealing wax was frequently formed of beeswax and the greenish-yellow Venetian turpentine extracted from the deciduous European Larch (*Larix decidua*), also known as the Common Larch.

COLOURED SEALS

Initially, wax seals were uncoloured; later, colours such as vermilion were added to create imperial red seals. Additionally, a range of other coloured waxes was developed to indicate the varying status of documents.

WAX MESSAGES

The Romans had an ingenious way to send written messages. Hinged wooden writing tablets coated with beeswax were written on by the sender. The recipient erased the message and wrote a reply.

WAX CANDLES

Beeswax candles burn cleanly, without creating smoke or soot, and are especially preferred for use in churches.

MILITARY BEARING

Beeswax was widely used in moustache creams in Victorian and Edwardian times to impart stiffness and a firm, military bearing.

MUSICAL VALUE

Tambourine surfaces are waxed to extend their lives, especially if the player is employing a 'thumb-roll' technique.

LONGBOWS

The legendary English longbow, with a flight distance of 220 metres (240 yards), has bowstrings waxed to reduce air friction; the wax also gave protection from excessive moisture.

DOMESTIC VALUE

A coating of beeswax keeps granite worktops shiny and bright. Apply a light coating of wax, allow to dry, and wipe and polish with a piece of suede cloth.

CHAPTER SEVEN

Proverbs, Limericks & Verse

BEE PROVERBS & SAYINGS

History has peppered many languages with proverbs and sayings about bees; some are amusing, a few are reflective, and others are perceptive of our close association with bees and how our thoughts and actions are mirrored by them.

'A drop of honey will not sweeten the ocean'

'A still bee gathers no honey'

'Dead bees maketh no honey'

'As busy as a bee'

'Catch more with honey than vinegar'

'Every bee's honey is sweet'

'He has a bee in his bonnet'

'Honey sometimes turns sour'

'Honey you swallow, gall you spit'

'Honey young, wine old'

'If you are too sweet, the bees will eat you'

'If you love honey, don't fear the sting'

'Make honey out of yourself and the flies will devour you'

'One bee is better than a handful of flies'

'Swine, women and bees cannot be turned'

'The diligence of the hive produces the wealth of honey'

'What is good for the swarm is not good for the bee'

'Where there is honey, the bears will come uninvited'

'Where there is honey, there are bees'

'Bees that have honey in their mouths have stings in their tails'

'No bees, no honey, No work, no money'

'When bees are old they yield no honey'

'To have your head full of bees'

BEE LIMERICKS

These distinctively witty and humorous nonsense poems were known in England during the early years of the 1700s. They were later popularized by Edward Lear and became an amusing form of poetry. They were considered to be a rich and original form of folklore.

There was an Old Man in a tree,
Who was horribly bored by a bee.
When they said, 'Does it buzz?'
He replied, 'Yes, it does,
It's a regular brute of a bee!'

There was an Old Man of Kilkenny,
Who never had more than a penny;
He spent all that money,
On onions and honey,
That wayward Old Man of Kilkenny.

Edward Lear

There was an Old Person of Dover,
Who rushed through a field of blue Clover;
But some very large bees,
Stung his nose and his knees,
So he very soon went back to Dover.

EDWARD LEAR (1812–1888)
Book of Nonsense

An unusual bee-keeper's garden and hives

There was an old man of St Bees,
Who was horribly stung by a Wasp.
When asked, 'does it hurt?'
He replied, 'No, it doesn't;
I'm so glad it wasn't a hornet.'

W.S. GILBERT (1836–1911)
[AFTER EDWARD LEAR]

Concerning the bees and the flowers,
In the fields and the gardens and bowers,
You will note at a glance,
That their ways of romance,
Haven't any resemblance to ours.

ANONYMOUS

BEES IN VERSE

Bees and honey have long featured in our literary heritage, for example in romantic and bucolic verses we may have first heard many years ago. Despite the age of these verses, they still have the ability to make us pause for thought and reflection.

There's a whisper down the field where the year has
  shot her yield,
And the ricks stand grey to the sun,
  Singing:
'Over then, come over, for the bee has quit the clover,
  And your English summer's done.'

RUDYARD KIPLING (1865–1936)
The Long Trail

And many a thought did I build up on thought,
As the wild bee hangs cell to cell.

ROBERT BROWNING (1812–1889)
Paracelsus

The moan of doves in immemorial elms,
And murmuring of innumerable bees.

ALFRED TENNYSON (1809–1892)
Come Down, O Maid

The bee goes singing to her groom,
Drunken and overbold.

ROBERT BROWNING (1812–1889)
Popularity

The moth's kiss first!
Kiss me as if you made believe
You were not sure, this eve,
How my face, your flower, had pursued,
Its petals up …
The bee's kiss, now!

ROBERT BROWNING (1812–1889)
In a Gondola

Love, in my bosom, like a bee,
Doth suck his sweet:
Now with his wings he plays with me,
Now with his feet.

THOMAS LODGE (1558–1625)
Love, In My Bosom

❧

While the bee with honied thigh,
That at her flowery work doth sing.

JOHN MILTON (1608–1674)
Il Penserosa

❧

Star that bringest home the bee,
And settest the weary labourer free.

THOMAS CAMPBELL (1771–1844)
Song to the Evening Star

❧

The poison of the Honey Bee
Is the Artist's Jealousy

WILLIAM BLAKE (1757–1827)
Gnomic Verses

There's a little short gentleman,
That wears the yellow trews,
A dirk below his doublet,
For Sticking in his foes,
Yet in a singing posture,
Where 'er you do him see,
And if you offer violence,
He'll stab his dirk in thee.

EARLY 18TH-CENTURY VERSE

How doth the little busy bee
Improve each shining hour,
And gather honey all the day
From every opening flower.

REV. ISAAC WATTS
(1674–1748)
Against Idleness and Mischief

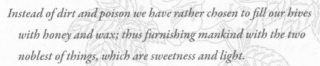

Instead of dirt and poison we have rather chosen to fill our hives
with honey and wax; thus furnishing mankind with the two
noblest of things, which are sweetness and light.

JONATHAN SWIFT (1667–1745)
Battle of the Books

INDEX

First published in the UK in 2011 by Green Books Ltd.

Reprinted in 2014 by
Green Books
PO Box 145, Cambridge, CB4 1GQ, England
www.greenbooks.co.uk
+44 (0)1223 302 041
Green Books is an imprint of UIT Cambridge Ltd.

The right of David Squire to be identified as the
author of this work has been asserted by him in
accordance with the Copyright, Designs and
Patents Act 1988.

The author wishes to thank
E. H. Thorne (Beehives) Limited for reference
material used from their catalogue.

DESIGNER
Glyn Bridgewater

ISBN 978 0 85784 024 0 (hardback)
ISBN 978 0 85784 216 9 (ebook – ePub)
ISBN 978 0 85784 039 4 (ebook – pdf)
Also available for Kindle.

10 9 8 7 6 5 4 3 2